THIS BOOK BELONGS TO

_____

MAKER

D1401385

Make:
ELECTRICITY FOR YOUNG MAKERS

FUN AND EASY DO-IT-YOURSELF PROJECTS

MARC DE VINCK ★ ART BY MIKE GRAY

MAKER MEDIA

San Francisco, CA

Electricity for Young Makers
Fun and Easy Do-It-Yourself Projects

Marc de Vinck

Art by Mike Gray

Published by Maker Media, Inc., 1160 Battery Street East, Suite 125, San Francisco, California 94111.

Maker Media books may be purchased for educational, business, or sales promotional use. Online editions are also available for most titles (safaribooksonline.com). For more information, contact our corporate/institutional sales department: 800-998-9938 or corporate@oreilly.com.

Publisher and Editor: Roger Stewart
Art Director, Designer, and Compositor: Jason Babler
Cover Designers: Jason Babler, Maureen Forys

Previously published as
Making a Penny-Powered Flashlight, a Flying LED Copter, and More!,
November 2016

Revision History for the This Edition

2017-04-15 First Release

978-1-680-45286-0

To my ~~biggest fan~~ wife Christa, my ~~beta testers~~ daughters Kaitlyn and Megan, and ~~most of~~ all of my editors both past and present for making my ~~ramblings~~ thoughts more acceptable for ~~books and stuff~~ print.

—Marc

For my Olivia, you are the best daughter in the world, and I love you more than anything.

—Mike

Marc de Vinck's latest challenge was to co-develop and co-teach in the Masters of Engineering in Technical Entrepreneurship (TE) program at Lehigh University in Pennsylvania. This award-winning TE program has graduated students from all over the world, instilling the importance of creativity and innovation in entrepreneurship. Prior to this, he was the Director of Product Development at *Make:* magazine and a member of the *Make:* Technical Advisory board. He has written numerous articles on do-it-yourself science and technology for *Make:*.

o———□———o

Mike Gray is an award-winning writer, cartoonist, animator, and director. He is also a proud father but, so far, has not won an award for that. You can peruse his portfolio online at pencilforhire.com.

Table of Contents

WELCOME TO THE WORLD OF ELECTRICITY AND ELECTRONICS!

Hi! I'm Makey!

Have you heard about the movement that's sweeping the globe? The one where people design, build, and share their creations and ideas with one another? It's called the Maker Movement! I was created to make, and I love to help people become Makers.

In this book, we're going to make stuff that will amaze and surprise you, make you see electricity in a different light, and, at the very least, put a smile on your face. That's because the projects are flat out fun! Along the way, you may even learn a few things that will help you in your science classes and maybe lead you to a career someday as an inventor, engineer, or scientist.

You don't have to read the whole book from cover to cover before you start making, but it's best to read each project all the way through before you start on it to be sure you know where you're going. I like to do the projects in order because they get a bit more complicated as the book goes along.

Let's get making!

BE SAFE!

I know. This isn't the amazing part, but following these safety rules is really important when it comes to electricity. And when you make anything that involves electricity, always think safety first!

NEVER EXPERIMENT WITH ELECTRICAL DEVICES THAT PLUG INTO THE WALL

Don't take anything apart that plugs into a wall outlet or a power strip even if it looks safe. It's not.

NEVER EXPERIMENT WITH LARGER BATTERIES

Our projects use household batteries, which are safe to handle. Larger batteries like the one in your car and rechargeable batteries pack a punch—and some toxic chemicals—so leave them alone!

THINK TWICE; ACT ONCE

Before you cut any wire or connect batteries, think about it. Is it a good idea? Can I undo what I'm about to do if it's wrong? Fortunately in this book almost everything can be remade if needed, so you don't have to worry too much about making mistakes!

PROTECT YOUR EYES

You should always wear eye protection during electrical experiments. Safety glasses are inexpensive, and you barely notice that you have them on, so go ahead and grab a pair and wear 'em. You only have one set of eyes!

NEVER EXPERIMENT WITH ELECTRICAL DEVICES THAT PLUG INTO THE WALL

OK, that was the first rule, but it's worth mentioning again. Yes, it's that serious!

MAKE A
PENNY-POWERED
FLASHLIGHT

Yes, you read that right. ⚡ We're going to power a flashlight with pennies. ⚡ No store-bought batteries here! ⚡ We'll be making our own power source with pennies and cardboard. ⚡ Did I mention that we're also building the flashlight? ⚡ We just need an LED and a few things you probably have around the house. ⚡ Prepare to illuminate!

TOOLBOX

SAFETY GLASSES

MEDIUM- OR FINE-GRIT SANDPAPER

SCISSORS

PAPER TOWELS

PAPER OR PLASTIC CUP

PENCIL

PARTS BIN

7 PENNIES, MADE IN 1983 OR LATER

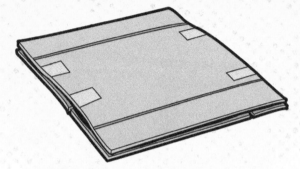

SCRAP CARDBOARD, ABOUT 6 INCHES SQUARE

VINEGAR, POUR INTO THE CUP

DUCT TAPE, 5-INCH-LONG PIECE

1 LED, RED

MAKE A PENNY-POWERED FLASHLIGHT

STEP 1

Gather seven pennies that were made in 1983 or later. You can tell when a penny was made by looking at the date on the front of it. Pennies made in 1983 and later have zinc in them, which is what we need for this project.

I HAVE ZINC ?!?

Did You Know?

The LED, or light emitting diode, was first invented in the 1960s, but the *idea* of a form of light that could be produced from silicon was conceived of way back in the early 1900s.

STEP 2

Put on your safety glasses and sand one side of each penny until it has a silver-looking finish. That's the zinc! It could take a while to sand away the copper (brown finish), but the zinc is under there.

STEP 3

Place a penny on a piece of cardboard and trace around it six different times. Cut out the six cardboard circles and put them in the cup of vinegar.

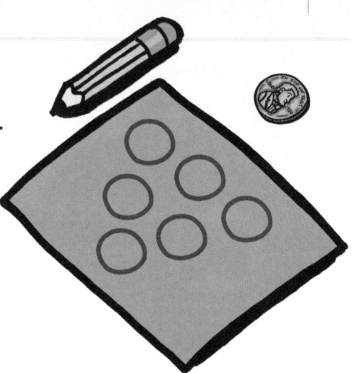

DO NOT DRINK

Once they are completely saturated with vinegar, take the cardboard circles out and use a paper towel to wipe off the excess vinegar.

STEP 4

Now let's build the battery. Cut a piece of duct tape about ½ inch wide and 5 inches long. Lay it flat on a surface—sticky side up. Place I sanded penny—zinc (silvery) side up—in the middle of the tape.

STEP 5

Place a cardboard circle on top of the penny. Place another penny—zinc side up—on top of the cardboard circle. Continue to do this until you've placed all pennies and cardboard circles in one stack. A penny with the zinc side facing up should be at the top of the stack.

MAKEY'S TIPS!

MAKE SURE NONE OF THE PENNIES ARE TOUCHING EACH OTHER. IF THEY TOUCH, IT WILL CREATE A SHORT CIRCUIT, AND THE BATTERY WON'T WORK. DON'T SQUEEZE THE STACK OF PENNIES EITHER. THE HARDER YOU SQUEEZE THE STACK, THE LOWER THE VOLTAGE THE BATTERY WILL PRODUCE. RED LEDS ARE BEST FOR THIS PROJECT BECAUSE THEY DON'T NEED MUCH VOLTAGE TO LIGHT UP, BUT IF YOU DON'T GENERATE ENOUGH VOLTAGE OR IF THERE'S A SHORT CIRCUIT, THE FLASHLIGHT WON'T WORK.

STEP 6

Bring one long end of the duct tape up along the side of the stack, wrap it across the top, and then down along the other side of the stack to the other side of the duct tape. The two long ends of the duct tape should meet at the bottom of the stack. Stick them together. Cut off excess tape after the long ends meet. The other two sides of the stack should remain open so we can attach the LED.

MAKE A PENNY-POWERED FLASHLIGHT

STEP 7

Take the LED and bend the wires, or leads, so the positive (+) lead—the longer one—can touch the zinc of the penny on the top of the stack. Bend the negative (−) lead—the shorter one—so it can touch the bottom penny's copper (brown) side. It should light up!

MAKEY'S TIPS!

IN OTHER ELECTRONIC PARTS, LONGER LEADS ARE ALMOST ALWAYS POSITIVE (+) AND THE SHORTER ONES ARE ALMOST ALWAYS NEGATIVE (−), JUST LIKE THE LEADS ON THIS LED.

Skill Builder

The penny battery is actually a voltaic pile that relies on the chemical reactions between the two electrodes—the copper and zinc—to create the energy that powers the LED. The LED should stay illuminated for hours, if not a whole day, depending on how long the electrolyte—in this case, the vinegar—lasts before it evaporates.

LATER...

TAKE IT FURTHER

You can increase the voltage by adding more pennies and cardboard circles to your stack. If you have access to a multimeter, use it to measure how much voltage your stack of pennies is creating and compare that to the label on the batteries in an inexpensive calculator. You can run two wires from your penny-pack to the battery compartment of the calculator. Be sure to note where the positive and negative wires need to attach, and test how much voltage your penny battery puts out before trying to use it!

What Is Electricity?

We just made electricity using chemistry... but what is electricity? Electricity is a form of energy caused by charged atomic particles. It can build up in one place or move from one place to another. When it's not moving it's called "static electricity," and when it is moving it's called "current electricity," or electrical current.

Have you ever walked across a carpet, then touched a doorknob and gotten a shock? That's static electricity! It happens because friction can knock electrons loose.

Some materials are hungrier for electrons than others, and as you move across the carpet you may build up a charge in one direction or the other. The metal doorknob acts as a conductor, allowing the built-up electrical charge to jump from you to the knob or from the knob to you. Either way, you feel it as a shock. That's electricity!

DIRECT CURRENT

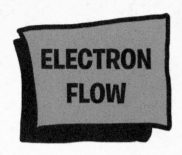

ELECTRON FLOW

ALTERNATING CURRENT

The other type of electricity, called electrical current, is what makes all of our appliances and gadgets work. It comes in two different varieties—AC and DC. They both involve the movement of an electric charge, but in slightly different ways.

Alternating current (AC) refers to an electrical charge that moves back and forth through circuits at a specific rate. It usually has much higher voltage than the batteries we are experimenting with and is much more dangerous. AC is used by anything that plugs into a wall outlet. Remember, don't experiment with alternating current from a wall outlet or power strip—it can seriously hurt you!

Direct current (DC) flows in one direction. Typically, anything that is powered by a battery or has a circuit board inside—such as your cell phone or laptop computer—uses DC. Although direct current can also be dangerous, it's pretty safe at the levels we're experimenting with using household batteries.

AC versus DC was once the subject of a great debate.

Electric power used to be distributed to our homes as direct current (DC), but eventually alternating current (AC) was found to be more efficient over long distances. We still use direct current (DC) in small electronic devices.

MAKE A FLYING LED COPTER

OK, so we made a battery out of pennies that powered an LED flashlight. ⚡ That was pretty cool. ⚡ How about we move on to something that not only lights up but also flies through the air?! ⚡ This is a two-part project, but I think you're ready for it. ⚡ Plus, we've got copters to launch at 1900 hours tonight!

TOOLBOX

SAFETY GLASSES

DUCT TAPE

SCISSORS

RULER

NEEDLENOSE PLIERS OR WIRE CUTTERS

PENCIL

PARTS BIN

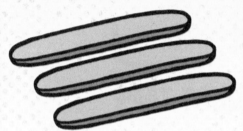

3 CRAFT STICKS

3 RUBBER BANDS, #33, THIN AND LONG

CR2032 BATTERY

STANDARD LED, ANY COLOR

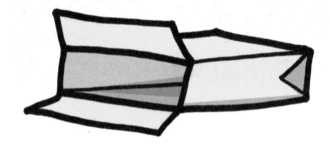

CEREAL BOX OR SIMILAR CARDBOARD

PART 1: BUILD THE LAUNCHER

STEP 1

Gather the materials for this part of the project. You'll need two of the three craft sticks, three #33 rubber bands, and some duct tape. Don't worry too much about the exact size of the rubber bands. Just make sure they are the thin type, stretchy and fairly long. #33 rubber bands are 1/8 inch wide and 3 1/2 inches long.

STEP 2

First let's make a chain of three rubber bands. Pinch the end of a rubber band to form a little hole about the size of dime. Next, partially thread another rubber band through the first rubber band, going around it, and back into itself. This makes a knot and a nice connection. Do the same with the third rubber band, creating a chain that is three rubber bands long.

STEP 3

Now let's attach the rubber band chain in the same way to a pair of craft sticks. Loop one end of the rubber band chain around a craft stick about ½ inch from the end. Then feed the other end of the rubber band chain through the loop. This should create a knot around the wooden craft stick. Do the same to the other end of the rubber band chain to another craft stick.

STEP 4

Next let's make a "V" shape with the craft sticks and secure them at the base with narrow strips of duct tape. Start by tearing a 4-inch-long piece of duct tape into three strips, each about ¼ inch wide. Wrap one piece of tape around one end of the two sticks to form a "V". The wide part of the "V" should be about 3 inches apart. Then, wrap a piece of tape vertically between the "V" at the base, followed by wrapping the third piece around the bottom of the "V" again.

STEP 5

And now for the last step. Carefully cut off one side of the middle rubber band in the chain without cutting the knots. We'll use this later to attach our copter. This will also make the launcher a bit stretchier.

That's it! We're done with the launcher.

PART 2: BUILD THE LED COPTER

STEP 1

Gather the electronic components for the build—the CR2032 battery and the LED. Let's take a look at the battery. The positive side is marked with a + sign. See it? On button cell batteries like this one, the negative side isn't usually marked, but the bottom must be negative, right? Remember when we made our own battery from pennies? We discovered that batteries have a positive end and a negative end, which allows the flow of electrons to power a circuit.

Now, let's check out the LED. It doesn't have a label for + or −, but it does have two wires, or leads, coming out of it. See how one is longer than the other? It's just like the LED we used in the mini-flashlight project. The longer wire is the positive (+) lead, and the shorter one is the negative (−) lead. Slide the LED onto the battery with the long lead touching the positive (+) side of the battery and the short lead touching the negative (−) side. It should light up.

I'M POSITIVE THE SHORT LEG IS NEGATIVE...

Did You Know?

An LED, or light-emitting diode, works when electrical voltage is applied to its leads. Inside the LED are two types of semiconductor materials—one that is negatively charged and one that is positively charged. When voltage is applied from the battery to these semiconductors, the negative and positive charge carriers are pushed together. When they combine, energy is released in the form of a photon (light particle). This process is called "electroluminescence."

STEP 2

Take the third, and last, craft stick and make a small triangular notch ½ inch from the end of the stick. You can make this notch with needlenose pliers or wire cutters by making small nibbles until it's perfect.

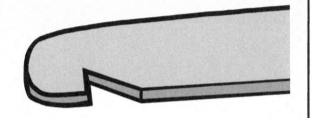

MAKEY'S TIPS!

BE SURE IT LOOKS JUST LIKE THE ONE IN THE PICTURE. IT NEEDS TO BE SHAPED LIKE A TRIANGLE, AND IT SHOULD HAVE A SLIGHT HOOK SHAPE TO IT. THE NOTCH SHOULD CUT MORE THAN ABOUT ⅓ OF THE WAY THROUGH THE CRAFT STICK.

STEP 3

Cut out two strips of cardboard—the kind cereal boxes are made of—that are 3/4 inch wide by 7 inches long. Now cut a slight taper into one end of each of these rectangles, about 1 inch from the end. This makes them a little narrower so you can easily tape them later.

STEP 4

Hold both pieces of cardboard together and make a 45-degree bend about 3 inches from the tapered end.

45° IS HALF A RIGHT ANGLE!

DOES THAT MEAN IT'S HALF RIGHT?

STEP 5

Tape the two strips of cardboard to opposite sides of the craft stick, right above the notch you cut earlier. Make sure the tape overlaps the cardboard and craft stick but doesn't cover the notch. The strips should bend away from the craft stick. These are the "copter blades."

MAKE A FLYING LED COPTER

STEP 6

There's just one more thing to do—attach the LED and battery to the copter. Slip the LED onto the battery so that the correct leads are touching the + and − sides. The LED should light up. Wrap one more piece of tape around the battery, LED, and craft stick sandwich behind the notch.

MAKEY'S TIPS!

THE BATTERY IS LARGER THAN THE CRAFT STICK. MAKE SURE IT OVERHANGS THE CRAFT STICK ON THE SIDE WITHOUT THE NOTCH. IF IT OVERHANGS BY THE SIDE WITH THE NOTCH, IT WILL BE HARD TO LAUNCH.

We now have a launcher and a glowing LED copter that's ready for action. All we need is a big space, preferably outside, and some darkness! Sure you can try some daytime flights, but where's the fun in that?!

Do you want to launch your copter? Of course you do! As all good pilots know, safety comes first, so *put on your safety glasses.* Now hook the notch at the front of the copter onto the center of the rubber band chain on the launcher. Extend your arms away from your body and face. Aim up and away from yourself and anyone nearby. Pull back on the craft stick of the copter, and let go!

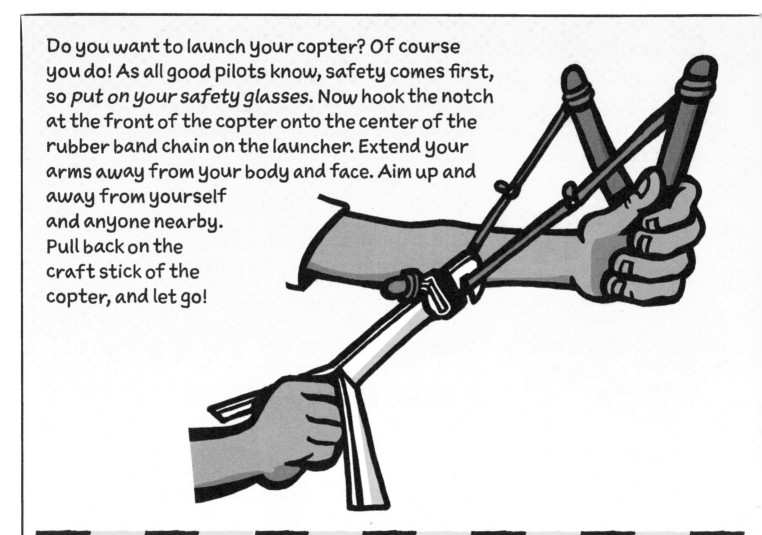

STAY SAFE!

Be sure to keep the copter away from your face and don't pull back too much. The copter should be launched from around your shoulder height and pointed away from you and toward the sky at about a 45-degree angle. Pull the copter back toward your waist. Be sure to wear safety glasses when launching your copter, and never point it at anyone or any animal!

MAKE A FLYING LED COPTER

TAKE IT FURTHER

Experiment with different wing shapes and LED configurations. Can you make your copter stay up longer, go further, or have different colored light patterns? This is also a great project to practice your long-exposure photography. Try taking a picture with a camera mounted on a tripod or stable surface. If you leave the camera shutter open for a long time, you should be able to capture a stream of light like a fireworks show.

What Is an Electrical Circuit?

In the copter project, when you connected the two wires of the LED to the button cell battery you created a circuit. But what *is* a circuit, and why was it necessary to know which side is positive (+) and which side is negative (−)?

A circuit is a path that allows electrons to flow. The path has to be made of a conductive material, which lets electrons easily leap from one atom to the next. But it also has to be a closed path, or loop.

LOOK OUT BELOW!

If the circuit is broken, or not a complete loop, the current won't flow. An on-off switch is simply a device that lets you temporarily break the circuit, making all the electrons stop. It also allows you to easily close the circuit again, which restarts the flow of electrons!

The batteries we use work by creating high potential energy at one end and low potential at the other through an electrochemical cell. The electromotive force of this cell is measured in volts (V).

I HAVE A LOT OF POTENTIAL!

HIGH POTENTIAL

LOW POTENTIAL

MAKE A
SPEAKER OUT
OF PAPER

In the last two projects we used electricity to create light. ⚡ How about we try to make sound with electricity? ⚡ In this project we'll see what amazing things happen when we bring electricity and magnets together.

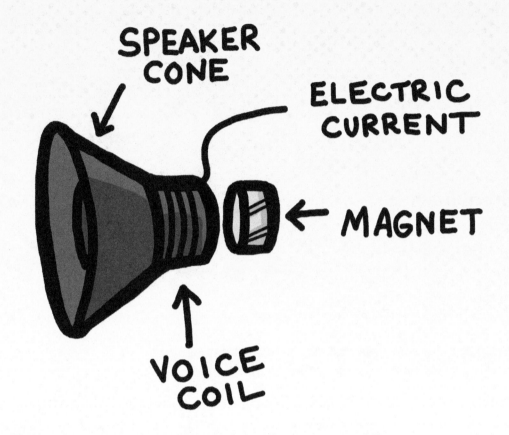

SPEAKER CONE

ELECTRIC CURRENT

MAGNET

VOICE COIL

HOW SPEAKERS WORK

Have you ever wondered how a speaker works? It's surprisingly simple! The two wires leading into the speaker carry a varying electrical signal to a wire coil, known as the "voice coil." This signal causes the electrons in the coil to move and create a magnetic field that attracts and repels a magnet, moving it up and down depending on which direction and how much electrical current is flowing through the coil. This movement is transferred to the cone-shaped part of the speaker. The vibrating cone moves the air in front of it, and our ears translate that into sound.

DID YOU KNOW?

Believe it or not, cells in your ear transform these vibrations back into electrical signals that your brain can understand. Electricity... it's everywhere. Even in your brain!

GETTING STARTED

Take a close look at the parts bin. You'll need a length of copper wire that has a very thin coating of nonconductive material as well as some strong ½-inch-diameter neodymium disc magnets. You should be able to find these things at a local hardware or electronics store. Last, look around to see if you can find a cheap pair of unwanted earbuds or headphones that's OK to take apart.

Note: The thickness of wire is measured by its gauge. For this project, you want wire that's around 30 AWG (American wire gauge). The electrical wire in your home is likely to be 14 AWG. Larger numbers correspond to smaller diameter wire. Go figure!

TOOLBOX

SAFETY GLASSES

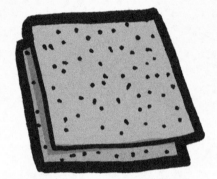

SMALL PIECE OF SANDPAPER

SCISSORS

HOT GLUE GUN OR CRAFT GLUE (OPTIONAL)

PARTS BIN

2 3X5-INCH INDEX CARDS

6 FEET OF MAGNET WIRE OR ENAMELED COPPER WIRE, 30 AWG

PAIR OF OLD EARBUDS OR HEADPHONES

CLEAR TAPE

3 NEODYMIUM DISC MAGNETS

STAY SAFE!

All magnets can be potentially dangerous. In this project we are using neodymium (pronounced nee-oh-DIM-ee-um) magnets, which are the strongest permanent magnets you can buy. Keep them away from small children, who might swallow them, and never allow them to crash together as they can easily shatter. It's easy to get pinched, and that's never fun!

PART 1: BUILD THE VOICE COIL

STEP 1

Stack the three neodymium magnets into a tower. Measure the height of the stack. Cut a 5-inch-long strip from one of the index cards that is the same width as the height of the stack of magnets.

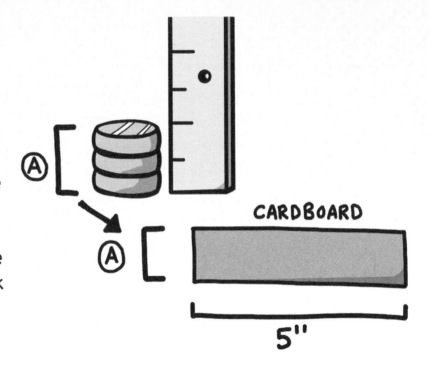

CARDBOARD

5"

STEP 2

Wrap the strip of paper three times around the stack of magnets and secure it at the end with a small piece of tape. Wrap a piece of tape—sticky side out—one to two times around the outside of the paper and magnet tower.

STEP 3

It's time to make the coil! Cut off a 6-foot length of magnet wire. Leave about 6 inches of the end of the wire free, and then wrap the rest around the magnet stack 50–60 times on top of the sticky tape. Leave another 6-inch tail of wire at the end.

STEP 4

Take an additional piece of tape and wrap it all the way around the coil—this time sticky side in—so it secures the wire in place.

STEP 5

Slide the stack of magnets and the paper surrounding them out from inside the wire coil. It should slip out pretty easily. You now have a cylinder of enameled copper wire sandwiched between strips of tape. You no longer need the strip of paper that you used to wrap the magnets. The voice coil should have a slightly bigger diameter than the stack of magnets.

STEP 6

To complete the voice coil, you need a way to connect those two free wires to an audio source. This is where that old pair of cheap headphones or earbuds come into the picture. Be sure you have permission before continuing! Cut off the jack (the plug end), leaving about 6 inches of the cable attached to it.

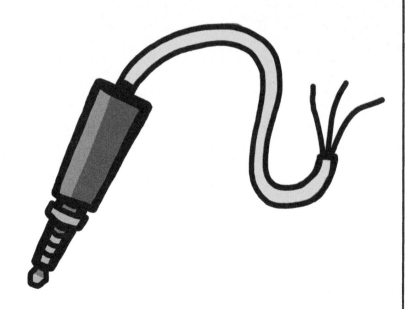

STEP 7

Strip about 1 inch of the plastic coating on the cut end of the headphone wire (the part with the jack). Inside you should see two or three smaller wires. It would be great if they were all color-coded the same way, but they aren't. In most cases you will want to use the yellow and green wires, or red and green, but you might have to play around with it to see what works.

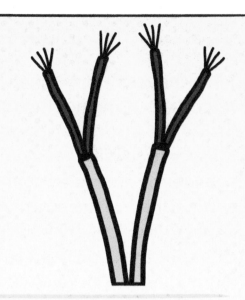

STEP 8

Use a small piece of sandpaper to remove about 1 inch of the enamel coating on the two ends of your voice coil. The coating makes it nonconductive. We want to make an electrical connection, so we need to clean it off to get to the bare copper wire.

STEP 9

Entwine the bare wires at the end of your voice coil around the bare wires at the end of the headphone jack. Don't make them too tight yet; you may need to experiment to find the right connection.

DON'T WORRY ABOUT WIRING UP THE WRONG TWO WIRES. THERE ISN'T MUCH ELECTRICITY COMING FROM A HEADPHONE JACK. JUST MAKE SURE ONE END OF THE VOICE COIL GOES TO ONE WIRE OF THE HEADPHONE WIRE, AND THE OTHER END OF THE VOICE COIL WIRES CONNECTS TO A DIFFERENT WIRE OF THE HEADPHONE JACK.

Okay, take a breath! We finished the voice coil. That was the hard part. Now we're going to make the speaker diaphragm. This is the part that will vibrate, like the cone in a regular speaker, and make sound. We're going to make it out of that other index card you've been wondering about.

PART 2: MAKE THE DIAPHRAGM

STEP 1

Mark the center of the index card. Don't worry about getting it perfectly in the center. Just make a small X marking the general location.

STEP 2

Now let's attach the voice coil to the center of the index card, using the X as a guide. The best way is to make a circle of hot glue on the card and place the circular edge of the voice coil on it. If you don't have a hot glue gun, you can use craft glue. (Craft glue takes a little longer to dry.) Be sure the coil is secured to the index card; it will be bouncing around when music plays. The better the connection between the coil and the index card, the better the sound will be.

STEP 3

Fold down the four corners of the index card, making legs so it can sit about an inch above the table or slightly higher than the stack of magnets.

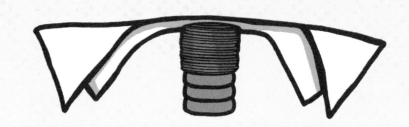

STEP 4

Place the neodymium magnets in a stack on a hard surface, such as a table.

STEP 5

Plug in the headphone jack into a sound source, such as an MP3 player, smartphone, or laptop computer. Select some music and press play.

STEP 6

Gently lower the voice coil down over the stack of magnets. Press down on the index card until you hear the music playing. Adjust the legs so the card and coil sit at the ideal height to produce the loudest sound. Did you just make that index card sing?!

NOT HEARING ANYTHING? TEST TO BE SURE THE WIRES ARE CONNECTED THE RIGHT WAY AND THAT THE CONNECTIONS ARE TIGHT. MAKE SURE THE MUSIC IS PLAYING. ADJUST THE VOLUME OF THE SPEAKER BY ADJUSTING THE LEGS ON THE PAPER DIAPHRAGM. IF THE COIL IS TOO HIGH OR TOO LOW, YOU WON'T GET ANY SOUND. FIND THE SWEET SPOT AND HAVE SOME FUN.

TAKE IT FURTHER

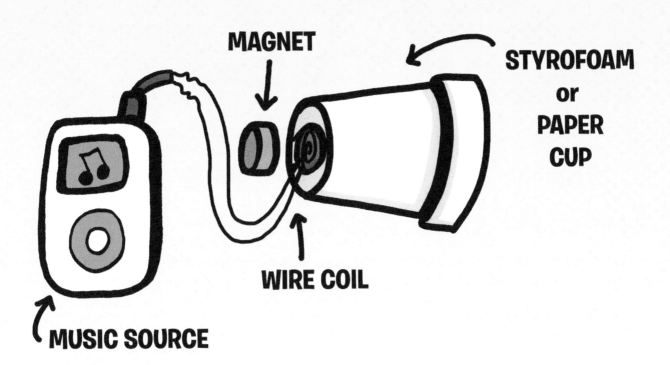

MAGNET

STYROFOAM or PAPER CUP

WIRE COIL

MUSIC SOURCE

The paper diaphragm doesn't have to be an index card. In fact, it doesn't even have to be paper. What else could you use, or how could you make it look more interesting?

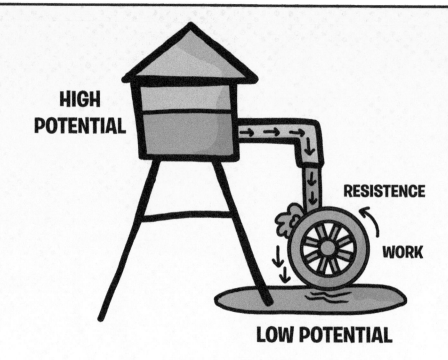

HIGH POTENTIAL

RESISTENCE

WORK

LOW POTENTIAL

More About Electricity

When electrons move between the atoms in a circuit, a current of electricity is created. It's this "flow of electrons" that we typically think of as electricity.

Electrical energy stored in a battery is like water stored in a water tower. When you give the energy in the battery a way to get out, it will flow in a direction that has less electrical energy. The "way out" is a conductive material such as copper wire.

BUT WHAT'S FLOWING IN AN ELECTRICAL CURRENT?

The answer is electrons. You might have learned about electrons in your science classes. They are the part of the atom that surround the nucleus.

Electrons can change their level in relation to the core of the atom, and they can even escape entirely. The electrons at the most outer levels of the atom have the least good grip and are easiest to knock loose.

Free electrons are what create the flow of electric charge. Electrons all carry a negative charge, and particles that have the same charge repel, or push each other away. Copper wire is a good conductor of electrical charge because the outer electrons in copper atoms can be knocked loose easily. The loose electron jumps to the next atom, and—because they repel—pushes another electron free. This sets up a chain reaction that moves the electrical charge through the wire.

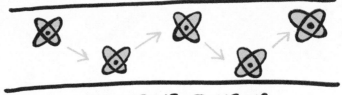

ELECTRONS JUMPING FROM ATOM TO ATOM

MAKE
TWO SIMPLE
MOTORS

Light. ⚡ Sound. ⚡ How about action? ⚡ Let's see if we can use electricity to make things move. ⚡ An electric motor is a machine that converts electrical energy into mechanical energy. ⚡ Products that use electric motors every day include toys, kitchen appliances, and even electric cars.

TOOLBOX

SAFETY GLASSES

NEEDLENOSE PLIERS (OPTIONAL)

So far we've made our own electricity with a battery, learned about the flow of electric current in circuits, and seen how electric fields interact with magnets. Now we're going to use all of those concepts to make a really simple motor called a "homopolar" (or 1-pole) motor.

PARTS BIN

1 6-INCH LENGTH OF SOLID COPPER WIRE

1 10-INCH LENGTH OF SOLID COPPER WIRE

1 AA BATTERY

1 SCREW, ABOUT 1 INCH IN LENGTH

1 NEODYMIUM DISC MAGNET

Skill Builder

The wire that you find at your local hobby shop comes in many different variations—from multistrand to solid core and from thick to thin. Thickness is measured as its gauge, or ga for short. For this project you'll need *bare* copper wire that is between 14ga and 22ga (or 14 AWG and 22 AWG). If your wire has a plastic coating, you can usually remove it fairly easily with pliers.

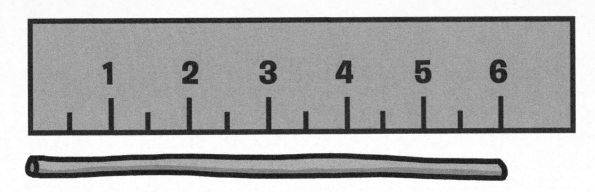

PART 1: BUILD SIMPLE MOTOR 1

All three of these motors are simple, but the first is far and away the simplest of all.

STEP 1

Cut a piece of wire about 6 inches long. If your wire has a plastic coating, remove about ¼ inch from each end.

STEP 2

Attach a neodymium magnet to the top (head) of your screw. The length of the screw isn't important, neither is the exact size of the magnet. Just be careful when working with magnets, especially neodymium magnets because they are really strong!

STEP 3

Holding the battery with the positive (+) end facing down, suspend the screw from it by gently placing the sharp tip of the screw against the positive part of the battery. It should hang there because of the magnetic field from the magnet on the other end of the screw.

Did You Know?

The first homopolar motor was invented back in the early 1800s by a physicist and chemist named Michael Faraday.

STEP 4

To get the motor spinning, hold one end of the wire on the negative (−) side of the battery and the other end on the edge of the magnet. Your motor will start to spin almost like magic. *Is it magic? No way! It's physics!*

I MUST POSSESS THIS MAGIC!

TO MAKE THE SCREW STAY CENTERED ON THE POSITIVE END OF THE BATTERY, TRY USING THE BLUNT TIP OF A PAIR OF NEEDLENOSE PLIERS TO MAKE A SLIGHT INDENTATION IN THE BATTERY TIP. THE METAL IS REALLY THIN, SO YOU DON'T NEED TO PUSH TOO HARD.

MAKEY'S TIPS!

Did You Know?

The magnet spins because of something called the "Lorentz (pronounced low-RENTS) force." When charged particles—in this case, electrons—enter the magnetic field at the point where the wire touches the magnet, a force that is perpendicular to both of them is created, which causes the spin.

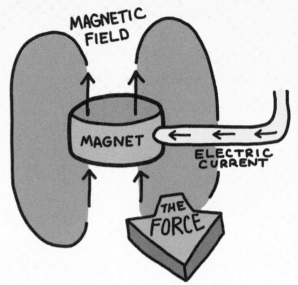

MAGNETIC FIELD

MAGNET

ELECTRIC CURRENT

THE FORCE

WHOA! I'M FEELING THE LORENTZ FORCE!

TRY THIS!

What happens when you switch the battery the other way around?

PART 2: BUILD SIMPLE MOTOR 2

The first motor we made worked really well, and was very simple. Now let's make something a bit more fun that still relies on the same basic principles.

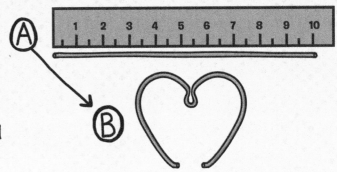

STEP 1

Bend a piece of bare copper wire about 10 inches long into the shape of a heart. You can use the pattern on the next page as a template. Be sure the wire doesn't have a plastic coating or it won't work. Also, be sure it's perfectly balanced. You can test its balance by placing the dip at the top of the wire heart onto your fingertip. If it doesn't balance on your finger, it won't balance on the battery.

STEP 2

Place a neodymium magnet at the bottom of the battery. This is the negative (–) terminal.

TEMPLATE FOR BENDING WIRE

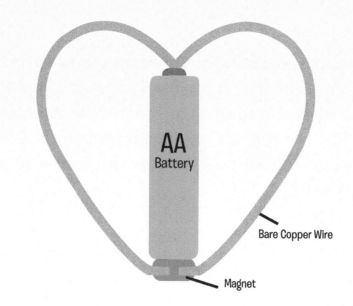

AA Battery

Bare Copper Wire

Magnet

STEP 3

Place the heart-shaped wire on the battery so that the positive end of the battery touches the center point of the heart, and the bottom ends of the wire touches the sides of the magnet. If your magnet is smaller than the battery, it's not a problem. Just bend the bottom of the heart so at least one of the wires touches the side of the magnet.

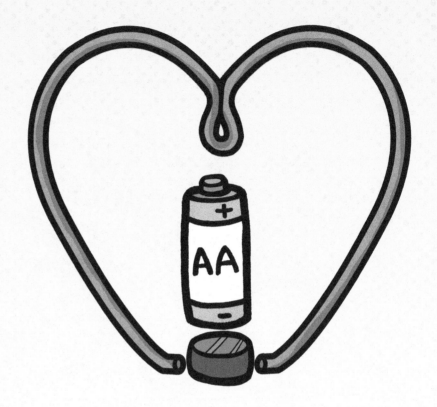

WHAT? IT'S NOT SPINNING? BE SURE AT LEAST ONE END OF THE WIRE AT THE BOTTOM TIP OF THE HEART IS MAKING CONTACT WITH THE EDGE OF THE MAGNETS AND NOTHING ELSE. IT SHOULD SPIN FREELY. ALSO, BE SURE YOUR BATTERY ISN'T DEAD!

MAKEY'S TIPS!

TAKE IT FURTHER

You can make almost any shape with a little time and help from your imagination. The tricky part is making sure that the majority of the weight is toward the bottom of your wire structure, otherwise it will be top heavy and fall over. The Makers back at the maker shed made one of yours truly!

How Do We Measure Electricity?

Electricity can be measured in these three ways:

★ CURRENT

Measured in amps (A), current measures the amount of charge in a circuit over a period of time. The symbol for current is I, which comes from the French *intensité de courant* (meaning "current intensity").

★ VOLTAGE

Measured in volts (V), voltage measures the force causing the electrons to travel. Stated differently, it measures how much potential energy exists between two points in a circuit.

★ RESISTANCE

Measured in ohms (OH-mms) (Ω), resistance measures how hard the electrons have to be forced to move through a circuit.

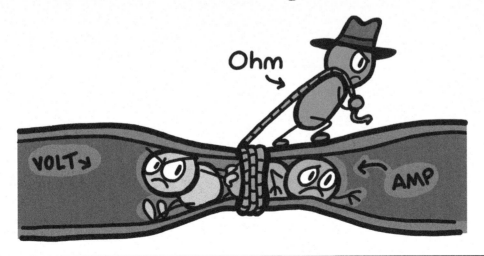

A discussion about ways to measure energy can't be complete without mentioning Ohm's Law—one of the most fundamental principles of electricity and electronics. Its name comes from Georg Ohm, a physicist and mathematician who sought to understand the relationship between voltage, current, and resistance.

The value of Ohm's Law is that when you have two known quantities, you can plug those into a formula to find the third, unknown, quantity. It is often expressed as

$$V = I * R$$

which means, voltage (V) measured in volts is equal to the current (I) measured in amps times the resistance (R) measured in ohms (Ω).

For example, $1V = 1A * 1\Omega$ means 1 volt is required to produce a current of 1 amp through a resistance of 1 ohm.

RESISTANCE IS FUTILE!

WELL, THAT'S WHERE YOU'RE WRONG!

MAKE TWO SIMPLE MOTORS

MAKE A
MORE ADVANCED
MOTOR

Now, let's make a more advanced motor that looks and works more like the kind we see everyday. ⚡ Most small electric motors use permanent magnets and some type of coil, or multiple coils, to convert electric current into an alternating electromagnetic field that can attract or repel the permanent magnets. ⚡ By carefully timing this alternating attraction and repelling, we can move the coil and create motion.

TOOLBOX

SAFETY GLASSES

FINE-GRIT SANDPAPER, 200 GRIT OR MORE

NEEDLENOSE PLIERS (OPTIONAL)

PARTS BIN

12 INCHES OF MAGNET WIRE OR ENAMEL-COATED WIRE (22 AWG)

2 2-INCH PIECES OF COPPER WIRE, OR 2 LARGE SAFETY PINS

1 AA BATTERY

DUCT TAPE

1 DISC MAGNET, STANDARD OR NEODYMIUM

MAKE A MORE ADVANCED MOTOR

STEP 1

Wrap the magnet wire around the AA battery 10 times, leaving about a 2-inch tail on either end.

STEP 2

Remove the coil from the battery and carefully wrap the tails around the coil 2–3 times to hold it all together. Make sure the ends of the wire stick straight out on opposite sides of the coil.

STEP 3

Hold the coil vertically in one hand, and gently sand off the enamel coating from the **top sides** of the two tails of the coil. Hold the coil vertically in one hand, and gently sand off the enamel coating. Be sure not to sand both top and bottom of the wire, we only want half of the wire sanded. Both tails should still have enamel on the bottom sides. Take your time sanding, moving the sandpaper in one direction only, and only use *gentle* pressure. The enamel coating is really thin; it won't take long to remove.

GO IN ONE DIRECTION

IF YOU CAN USE THE EDGE OF A PIECE OF SCRAP WOOD TO SAND AGAINST, IT ALLOWS YOUR COIL TO BE HELD UPRIGHT A LOT EASIER WHILE SANDING. JUST MAKE SURE NOT TO USE THE EDGE OF A GOOD PIECE OF FURNITURE. THE SANDING WOULD RUIN IT!

MAKEY'S TIPS!

MAKE A MORE ADVANCED MOTOR

STEP 4

Now we need to create the armature that will hold the coil above the magnet. Take two 2-inch-long pieces of copper wire and make a small loop at both ends of each wire. If the wire has a plastic coating, strip the coating off on both ends.

Tape the wires to each end of the battery. You can take a strip of tape and wrap it the long way around the battery to hold the armature in place.

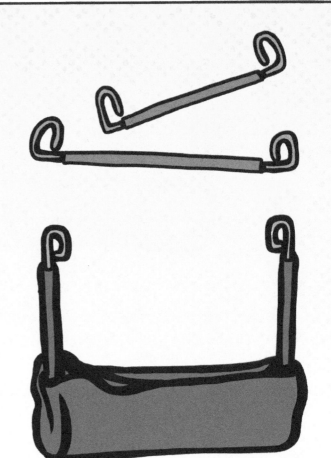

MAKEY'S TIPS!

IF YOU DON'T HAVE COPPER WIRE, YOU CAN USE TWO MEDIUM-SIZED, ALL-METAL SAFETY PINS. TAPE THEM TO THE BATTERY JUST AS YOU WOULD HAVE THE COPPER WIRES. TAPE THEM WITH THE CLOSURE PART OF THE PIN AGAINST THE BATTERY.

STEP 5

Put the battery on a flat surface and add the magnet to the center of the battery. Place the coil above the magnet, using the two loops to hold it.

Give the coil a spin, and you should see it start to spin... fast!

IF THE COIL ISN'T SPINNING, CHECK TO MAKE SURE YOU HAVE A FRESH BATTERY. ALSO MAKE SURE THE COIL IS AS CLOSE TO THE MAGNET AS POSSIBLE AND STAYS CENTERED BETWEEN THE SAFETY PINS. BALANCE THE COIL AS WELL AS YOU CAN, KEEPING THE TAILS PERFECTLY STRAIGHT.

MAKEY'S TIPS!

Did You Know?

This type of motor works by allowing the electricity to flow to the coil *only* when an electrical connection can be made, which happens where you sanded off the enamel coating. This connection makes the coil attract or repel the permanent magnet, making it turn. Once it turns about half way, the electrical connection is lost because of the remaining enamel coating on the wire. With the circuit broken, the coil continues to spin because of its momentum but slows down due to friction. It becomes energized again after half a turn and resumes spinning. You may see the coil swing back and forth at first, which really illustrates this concept; but eventually it has enough momentum to fully spin around and around.

WHAT'S NEXT?

Congratulations! If you finished the projects in this book, you're now officially a Maker like me! You also have a foundational understanding of basic electricity and electronics. Experiment on your own. Try out modifications of the projects in this book or come up with new projects of your own. If you'd like to share your projects with other Makers, visit our website and tell us about it at makezine.com/contribute.

If you want to find more projects, you can watch videos online, and you can visit your local library or bookstore to find more books on the subject. The book *Make: Electronics* by Charles Platt is a good place to learn about building circuits with electronic components.

Look for a Maker Faire in your area to find out what other enthusiastic Makers are doing, and consider joining a makerspace in your community to learn from experienced Makers who are happy to share their knowledge with you!

Maker Faire ®

Maker Faire is the Greatest Show (and Tell) on Earth. It's a family-friendly festival of invention, creativity, and resourcefulness and a celebration of the Maker movement.

Maker Faires are held all over the world and are gatherings of people who love to share what they are doing. **Visit a Maker Faire in your area.** You'll meet other Makers like you and experience some of the amazing things they have created. You may even want to show off your own projects at a Maker Faire someday.

Find us online at MakerFaire.com to learn more about making and to find a Faire near you!

ALSO AVAILABLE

Learn how to use workshop tools while making fun and useful woodworking projects. Build valuable skills and learn the secrets of good craftsmanship.

★ Make a personalized wizard's wand for cosplay fun

★ Build a handy charging station for your cellphone

★ Construct a sturdy box with a secret compartment

★ Learn how the pros do it — stay safe and work smart

BE A MAKER!　　$12.99